Contract Negotiation Book!

Laurel D. Malvern

Copyright and Legal Disclaimer

© 2024 by Laurel D. Malvern. All rights reserved.

No part of this publication may be reproduced, distributed, or transmitted in any form or by any means, including photocopying, recording, or other electronic or mechanical methods, without the prior written permission of the publisher, except in the case of brief quotations embodied in critical reviews and certain other noncommercial uses permitted by copyright law.

This book is intended to provide general information and guidance on the topic of contract negotiation. The author and publisher have made every effort to ensure the accuracy of the information presented herein. However, the author and publisher do not warrant or represent that the information contained in this book is complete, accurate, or suitable for any particular purpose. Readers are advised to consult with legal, financial, or other professionals before making any decisions or taking any actions based on the information provided in this book.

The techniques, strategies, and examples discussed in this book are intended to be illustrative and educational. The author and publisher refuse any liability for any loss or damage resulting from reliance on the information contained in this book. Readers are solely responsible for their own actions and decisions in contract negotiation and related matters.

Chapter-Specific Legal Disclaimers:

Chapter: Fundamentals of Contract Negotiation

The information provided in this chapter is for general educational purposes only and should not be construed as legal advice. Readers are encouraged to seek the advice of qualified legal professionals regarding specific contract negotiation issues.

Chapter: Contract Negotiation in How-To Home Improvements
The techniques and recommendations discussed in this chapter are based on industry practices and common principles of contract negotiation. Readers should conduct thorough research and consult with relevant experts before entering contracts related to home improvement projects.

Chapter: Contract Negotiation in Engineering Construction
The information provided in this chapter is intended to offer insights into the negotiation process in the engineering construction industry. Readers are advised to seek the guidance of experienced professionals, such as engineers and construction lawyers, when negotiating complex construction contracts.

Chapter: Contract Negotiation in Civil Construction
The strategies and considerations outlined in this chapter are designed to assist readers in navigating contract negotiations in the civil construction sector. However, readers should be aware that legal requirements and industry standards may vary depending on district and project specifics.

Chapter: Contract Negotiation in Business
This chapter provides general guidance on negotiating contracts in various business contexts. Readers are encouraged to seek tailored advice from legal and financial professionals to address specific contractual issues and business objectives.

Chapter: Contract Negotiation in Administrative Law
The information presented in this chapter is intended to offer an overview of contract negotiation within the framework of administrative law. Readers should consult with qualified legal advisors to ensure compliance with relevant regulations and administrative procedures.

By accessing and using the information contained in this book, readers agree to indemnify and hold harmless the author and publisher from any claims, damages, or liabilities arising out of or related to the use of such information. This disclaimer applies to all chapters and sections of the book.

"Contract Negotiation Book"

- Chapter: Fundamentals of Contract Negotiation 8
- Chapter: Importance of Preparation: Research, Goal Setting, and Establishing Negotiation Parameters 11
- Chapter: Effective Communication Strategies: Active Listening, Asking the Right Questions, and Building Rapport 14
- Chapter: Contract Negotiation in How-To Home Improvements 17
- Chapter: Identifying Negotiation Priorities: Quality Standards, Timelines, and Budget Constraints 20
- Chapter: Negotiating with Contractors and Subcontractors: Scope of Work, Materials, and Warranties 23
- Chapter: Handling Disputes and Managing Changes Effectively 26
- Chapter: Contract Negotiation in Engineering Construction 29
- Chapter: Negotiating Complex Project Specifications 32
- Chapter: Balancing Cost and Quality in Engineering Construction 35
- Chapter: Mitigating Legal Risks and Ensuring Compliance 38
- Chapter: Contract Negotiation in Civil Construction 41
- Chapter: Negotiating with Government Agencies and Stakeholders 44
- Chapter: Addressing Environmental Concerns 47

Chapter: Navigating Public-Private Partnerships and Funding Arrangements 50

Chapter: Contract Negotiation in Business 53

Chapter: Negotiating Sales and Procurement Contracts 57

Chapter: Drafting Effective Service Agreements 60

Chapter: Managing Intellectual Property Rights and Confidentiality Provisions 63

Chapter: Contract Negotiation in Administrative Law 66

Chapter: Negotiating Contracts with Regulatory Bodies 70

Chapter: Addressing Legal Challenges and Regulatory Uncertainties 74

Chapter: Ensuring Transparency and Accountability in Government Contracts and Public-Private Partnerships 78

Conclusion: Mastering the Art of Contract Negotiation 82

Chapter: Reflecting on Personal Growth and Professional Development in Negotiation Skills 85

Chapter: Embracing Lifelong Learning and Adaptation in the Industry 89

Chapter: Upholding Ethical Conduct, Mutual Respect, and Win-Win Outcomes in Contract Negotiation 92

Chapter: Fundamentals of Contract Negotiation

Negotiating contracts is a fundamental skill that lies at the heart of every business transaction. In this chapter, we will delve into the foundational principles of contract negotiation, equipping you with the essential knowledge needed to navigate this intricate process effectively.

Understanding Key Terms:

Contracts are legal agreements that outline the rights, obligations, and responsibilities of the parties involved. Before diving into negotiations, it's crucial to understand some key terms commonly used in contract discussions:

Offer: An offer is a proposal made by one party to another, expressing an intention to enter a contract under specific terms.

Acceptance: Acceptance occurs when the offeree agrees to the terms of the offer, forming a binding contract between the parties.

Consideration: Consideration refers to something of value exchanged between the parties, such as money, goods, or services, which forms the basis of the contract.

Mutual Assent: Also known as "meeting of the minds," mutual assent occurs when both parties agree to the terms of the contract without coercion or misunderstanding.

Legal Capacity: Each party entering a contract must have the legal capacity to do so, meaning they must be of sound mind and legal age.

Parties Involved:

Contracts typically involve two or more parties, each with specific roles and responsibilities. These parties may include:

Offeror: The party making the offer or proposal.
Offeree: The party to whom the offer is made.
Promisor: The party making a promise in the contract.
Promisee: The party to whom the promise is made.
Third Parties: Individuals or entities not directly involved in the contract but who may be affected by its terms.
Understanding the roles of each party is essential for clarifying expectations and ensuring that all parties agree.

Legal Considerations:

Contract negotiation is not merely a matter of reaching a verbal agreement; it also involves adherence to legal principles and considerations. Here are some important legal factors to keep in mind:

Legal Requirements: Contracts must meet certain legal requirements to be enforceable, including offer, acceptance, consideration, legal capacity, and lawful purpose.

Statute of Frauds: Some contracts must be in writing to be enforceable, as mandated by the Statute of Frauds. These include contracts involving real estate, sales of goods over a certain value, and agreements that cannot be performed within one year.

Good Faith: Parties negotiating a contract are expected to act in good faith, meaning they must be honest and fair in their dealings and not engage in deceptive or unethical practices.

Contractual Terms: The terms of a contract should be clear, unambiguous, and specific to avoid confusion or disputes later.

By grasping these fundamental concepts of contract negotiation, you lay a solid groundwork for successful negotiations in any industry or context. In the following chapters, we will delve deeper into industry-specific negotiation techniques and strategies.

Chapter: Importance of Preparation: Research, Goal Setting, and Establishing Negotiation Parameters

Preparation is the cornerstone of successful contract negotiation. In this chapter, we will explore the critical importance of thorough preparation, including conducting research, setting clear goals, and establishing negotiation parameters. By investing time and effort in preparation, negotiators can enhance their leverage, anticipate challenges, and maximize the likelihood of achieving favorable outcomes.

Conducting Research:

Understanding the Industry Landscape: Begin by researching the industry in which the contract negotiation will take place. Familiarize yourself with industry trends, market dynamics, and the competitive landscape. This knowledge will provide valuable context and insights into prevailing norms and practices.

Gathering Information about the Counterparty: Research the counterparties involved in the negotiation, including their background, reputation, and past negotiation history. Understanding their priorities, preferences, and potential areas of flexibility can inform your negotiation strategy and help build rapport during discussions.

Analyzing Legal and Regulatory Requirements: Identify any legal or regulatory requirements that may impact the negotiation process or the terms of the contract. Consult legal experts to ensure compliance with relevant laws and regulations, reducing the risk of disputes or legal challenges down the line.

Setting Clear Goals:

Defining Objectives and Priorities: Clearly define your objectives and priorities for the negotiation. What are the key outcomes you hope to achieve? What concessions are you willing to make, and what are your non-negotiables? Setting clear goals provides focus and direction, guiding your decision-making throughout the negotiation process.

Establishing BATNA (Best Alternative to a Negotiated Agreement): Identify your BATNA — the best alternative course of action available if the negotiation does not result in a satisfactory agreement. Knowing your BATNA empowers you to assess proposals objectively and determine when to walk away from a deal that does not meet your objectives.

Creating SMART Goals: Ensure that your goals are Specific, Measurable, Achievable, Relevant, and Time-bound (SMART). By setting SMART goals, you create a roadmap for success and measure progress effectively during negotiations.

Establishing Negotiation Parameters:

Setting Boundaries and Limits: Determine your negotiation parameters, including acceptable ranges for key terms such as price, delivery timelines, and scope of work. Establishing clear boundaries communicates your expectations to the counterparty and provides a framework for constructive dialogue.

Identifying Flexibility and Trade-offs: Recognize areas where you may be willing to compromise or make concessions in exchange for favorable terms in other areas. Understanding your flexibility allows you to negotiate strategically and maximize value while maintaining alignment with your overall objectives.

Developing Communication and Decision-Making Protocols: Establish protocols for communication and decision-making within your negotiation team. Clarify roles and responsibilities, designate decision-makers, and define escalation procedures to streamline the negotiation process and avoid delays.

By prioritizing preparation, negotiators can position themselves for success and enhance their effectiveness at the negotiating table. In the subsequent chapters, we will explore advanced negotiation techniques and strategies tailored to specific industries and contexts.

Chapter: Effective Communication Strategies: Active Listening, Asking the Right Questions, and Building Rapport

Effective communication is the cornerstone of successful contract negotiation. In this chapter, we will delve into essential communication strategies that can enhance your negotiation skills, including active listening, asking the right questions, and building rapport. By mastering these techniques, negotiators can foster productive dialogue, uncover valuable insights, and strengthen relationships with counterparties.

Active Listening:

Give Your Full Attention: Actively listen to what the other party is saying without interrupting or formulating responses prematurely. Demonstrate your engagement by maintaining eye contact, nodding, and providing verbal cues to signal understanding.

Paraphrase and Clarify: Paraphrase the other party's statements to ensure accurate comprehension and demonstrate empathy. Clarify any points of ambiguity or uncertainty by asking open-ended questions and seeking clarification on key terms or concepts.

Empathize and Validate: Put yourself in the shoes of the other party and acknowledge their perspective, even if you disagree. Validate their concerns and emotions to foster trust and create a conducive environment for constructive dialogue.

Asking the Right Questions:

Open-ended Questions: Use open-ended questions to encourage the other party to share their thoughts, priorities, and preferences. Avoid leading questions that suggest a particular answer and instead allow room for exploration and discovery.

Probing for Understanding: Probe deeper to uncover underlying motivations, interests, and constraints. Ask follow-up questions to gain clarity on ambiguous points and uncover hidden opportunities for mutual value creation.

Clarifying Assumptions and Expectations: Challenge assumptions and clarify expectations to ensure alignment between parties. Ask questions to uncover any discrepancies or misunderstandings and address them proactively to prevent conflicts later in the negotiation process.

Building Rapport:

Establish Common Ground: Find common interests or experiences to establish rapport and build a connection with the other party. Shared experiences, hobbies, or professional backgrounds can serve as a foundation for building trust and rapport.

Demonstrate Authenticity: Be genuine and authentic in your interactions, avoiding scripted or overly formal language. Show genuine interest in the other party's perspective and demonstrate empathy and understanding.

Focus on Relationship Building: Prioritize relationship building over transactional outcomes, recognizing that long-term relationships are often more valuable than short-term gains. Invest time and effort in cultivating positive relationships with counterparties to lay the groundwork for future collaborations.

By mastering effective communication strategies such as active listening, asking the right questions, and building rapport, negotiators can create a conducive environment for constructive dialogue and collaboration. In the subsequent chapters, we will explore advanced negotiation techniques tailored to specific industries and contexts, building upon the foundation established through effective communication.

Chapter: Contract Negotiation in How-To Home Improvements

In this chapter, we'll delve into the fascinating world of contract negotiation within the realm of how-to home improvements. From renovations to remodeling projects, this industry presents unique challenges and opportunities for negotiation. We'll provide an overview of the industry, including current trends, common practices, and key stakeholders involved in the process.

Overview of the How-To Home Improvement Industry:

Current Trends: The how-to home improvement industry is constantly evolving, driven by changing consumer preferences, advancements in technology, and shifts in design aesthetics. Current trends include a focus on sustainable materials, smart home technologies, and multifunctional living spaces. Understanding these trends can help negotiators anticipate client preferences and adapt their proposals accordingly.

Common Practices: Home improvement projects often involve a series of common practices, including initial consultations, design planning, material selection, and construction or installation. Negotiators must navigate these phases effectively, addressing client concerns, managing expectations, and ensuring clear communication throughout the process.

Stakeholders: Several stakeholders play a crucial role in how-to home improvement projects, each with their own interests and objectives:

Homeowners: The primary clients seeking home improvement services. Their goals may include enhancing the aesthetics, functionality, or value of their property.

Contractors: Skilled professionals responsible for executing the renovation or remodeling work. Negotiations with contractors may involve discussing project scope, timelines, pricing, and quality standards.

Suppliers: Providers of building materials, fixtures, and appliances necessary for home improvement projects. Negotiations with suppliers may focus on pricing, delivery schedules, and product warranties.

Designers/Architects: Experts tasked with creating design concepts and blueprints for home improvement projects. Negotiations with designers/architects may involve discussing design preferences, project feasibility, and budget considerations.

Subcontractors: Specialized tradespeople hired by contractors to perform specific tasks within the project, such as plumbing, electrical work, or carpentry. Negotiations with subcontractors may involve subcontracting agreements, scheduling, and coordination with the main contractor.

Key Considerations in Contract Negotiation:

Scope of Work: Clearly define the scope of work to be performed, including specific tasks, materials, and deliverables. Negotiators should ensure alignment between client expectations and contractor capabilities to avoid misunderstandings or disputes later.

Budget and Pricing: Negotiate pricing and payment terms upfront, considering factors such as labor costs, material expenses, and profit margins. Establish a clear understanding of the project budget and any potential cost overruns to mitigate financial risks.

Timelines and Deadlines: Agree upon realistic timelines and deadlines for project completion, considering factors such as seasonal considerations, availability of materials, and subcontractor scheduling. Negotiators should build flexibility into the timeline to accommodate unforeseen delays or changes in scope.

Quality Standards: Establish quality standards and specifications for materials, workmanship, and finishes to ensure the desired level of craftsmanship. Negotiators should discuss quality control measures, inspections, and warranties to guarantee client satisfaction and project success.

By understanding the dynamics of the how-to home improvement industry, negotiators can navigate contract negotiations with confidence and achieve mutually beneficial outcomes for all stakeholders involved. In the following chapters, we will delve deeper into negotiation strategies and techniques tailored to specific aspects of home improvement projects.

Chapter: Identifying Negotiation Priorities: Quality Standards, Timelines, and Budget Constraints

In the realm of contract negotiation, identifying and prioritizing key objectives is essential for achieving successful outcomes. In this chapter, we will explore the importance of identifying negotiation priorities, with a focus on quality standards, timelines, and budget constraints in how-to home improvement projects. By understanding and effectively addressing these priorities, negotiators can lay the groundwork for successful project execution and client satisfaction.

Quality Standards:

Defining Quality Objectives: Quality standards encompass the level of craftsmanship, durability, and aesthetic appeal desired for the home improvement project. Negotiators should work closely with clients and stakeholders to define clear quality objectives, considering factors such as material selection, construction techniques, and finishing details.

Evaluating Trade-offs: In negotiation, it's crucial to balance quality considerations with other project constraints such as timelines and budget. Negotiators should assess trade-offs and explore creative solutions to achieve the desired level of quality within the available resources.

Quality Assurance Measures: Negotiators should establish mechanisms for quality assurance throughout the project lifecycle, including regular inspections, testing protocols, and adherence to industry standards. Clear communication and documentation of quality requirements are essential to ensure compliance and accountability.

Timelines:

Setting Realistic Deadlines: Timelines define the project schedule and milestones, outlining the sequence of activities from initiation to completion. Negotiators should collaborate with clients, contractors, and subcontractors to establish realistic deadlines that accommodate project complexity, resource availability, and external factors such as weather conditions.

Mitigating Schedule Risks: Negotiators should identify potential schedule risks and develop contingency plans to mitigate delays and disruptions. Strategies may include buffer periods in the project timeline, proactive communication to address issues promptly, and leveraging technology for efficient project management.

Progress Monitoring and Communication: Effective communication and progress monitoring are essential for keeping the project on track and addressing deviations from the schedule promptly. Negotiators should implement regular check-ins, status updates, and milestone reviews to ensure alignment between stakeholders and promote transparency throughout the project lifecycle.

Budget Constraints:

Establishing Budget Parameters: Budget constraints define the financial parameters of the project, including total costs, funding sources, and cost allocation for various project components. Negotiators should work collaboratively with clients to establish a realistic budget that aligns with project objectives and financial capabilities.

Cost-saving Opportunities: Negotiators should explore cost-saving opportunities without compromising project quality or integrity. Strategies may include value engineering, bulk purchasing discounts, and alternative material selections to optimize costs while meeting project requirements.

Managing Change Orders: Change orders may arise during the project, requiring adjustments to the original scope, timeline, or budget. Negotiators should establish clear protocols for managing change orders, including approval processes, cost implications, and documentation requirements, to minimize disruption and ensure project continuity.

By identifying and prioritizing negotiation priorities such as quality standards, timelines, and budget constraints, negotiators can effectively manage project expectations and achieve successful outcomes in how-to home improvement projects. In the subsequent chapters, we will explore advanced negotiation strategies for addressing specific challenges and scenarios in the home improvement industry.

Chapter: Negotiating with Contractors and Subcontractors: Scope of Work, Materials, and Warranties

Negotiating with contractors and subcontractors is a critical aspect of how-to home improvement projects. In this chapter, we will explore the key considerations and strategies for negotiating the scope of work, materials, and warranties with these essential stakeholders. By understanding their roles, addressing their concerns, and establishing clear expectations, negotiators can ensure smooth project execution and client satisfaction.

Understanding Contractor and Subcontractor Roles:

Contractors: Contractors are responsible for overseeing the overall execution of the home improvement project. They coordinate with subcontractors, manage project timelines, and ensure compliance with quality standards and building codes. Negotiations with contractors typically involve discussing the scope of work, pricing, timelines, and project management responsibilities.

Subcontractors: Subcontractors are specialized tradespeople hired by the main contractor to perform specific tasks within the project, such as plumbing, electrical work, or carpentry. Negotiations with subcontractors focus on scope, pricing, scheduling, and quality standards for their respective trades.

Negotiating the Scope of Work:

Defining Project Objectives: Clearly define the objectives and deliverables for the home improvement project, including specific tasks, milestones, and quality standards. Negotiators should collaborate with contractors and subcontractors to develop a comprehensive scope of work that aligns with client expectations and project requirements.

Addressing Scope Creep: Anticipate potential scope creep, which refers to uncontrolled expansion of project scope beyond the original agreement. Negotiators should establish change order procedures and protocols for addressing scope changes, including approval processes, cost implications, and schedule adjustments.

Materials Selection and Procurement:

Specifying Materials and Finishes: Negotiators should specify materials, finishes, and fixtures to be used in the home improvement project, considering factors such as durability, aesthetics, and budget constraints. Clear communication and documentation of material specifications are essential to avoid misunderstandings or disputes during construction.

Negotiating Pricing and Supplier Contracts: Negotiate pricing and terms with suppliers for materials and fixtures, leveraging bulk purchasing discounts and establishing favorable payment terms. Negotiators should review supplier contracts carefully, paying attention to warranties, delivery schedules, and dispute resolution mechanisms.

Warranties and Guarantees:

Ensuring Quality Assurance: Negotiate warranties and guarantees with contractors and subcontractors to ensure quality assurance and accountability for their workmanship. Warranties may cover defects in materials or workmanship for a specified period, providing clients with peace of mind and recourse in case of issues.

Reviewing Warranty Terms: Carefully review warranty terms and conditions, including coverage limitations, exclusions, and procedures for making warranty claims. Negotiators should seek to negotiate favorable warranty terms that protect the client's interests while being fair and reasonable to contractors and subcontractors.

Conclusion:

Negotiating with contractors and subcontractors requires careful planning, effective communication, and attention to detail. By addressing the scope of work, materials, and warranties in negotiations, negotiators can lay the foundation for successful project execution and client satisfaction in how-to home improvement projects. In the following chapters, we will explore additional negotiation strategies and techniques for managing project risks and achieving optimal outcomes.

Chapter: Handling Disputes and Managing Changes Effectively

In the dynamic environment of how-to home improvement projects, disputes and changes are inevitable. In this chapter, we will explore strategies for handling disputes and managing changes effectively, ensuring that negotiations remain productive, and projects stay on track.

Identifying Common Sources of Disputes:

Scope Changes: Changes to the project scope, whether due to client requests, unforeseen issues, or design modifications, can lead to disputes over additional costs, delays, and responsibilities.

Quality Concerns: Disputes may arise if there are discrepancies between the expected quality standards and the delivered workmanship, materials, or finishes.

Payment Disputes: Issues related to payment, invoicing, and billing discrepancies can strain the client-contractor relationship and lead to disputes over outstanding payments or project costs.

Communication Breakdowns: Miscommunication or lack of clarity in project communications can contribute to misunderstandings, delays, and conflicts between parties.

Strategies for Handling Disputes:

Open Communication: Foster open and transparent communication between all parties involved in the project. Encourage regular check-ins, status updates, and project meetings to address concerns proactively and prevent misunderstandings from escalating into disputes.

Mediation and Arbitration: In the event of a dispute, consider alternative dispute resolution methods such as mediation or arbitration. These processes provide a neutral forum for parties to resolve conflicts amicably with the assistance of a mediator or arbitrator.

Document Everything: Maintain detailed records of all project-related communications, agreements, and changes. Documenting discussions, decisions, and approvals can help clarify responsibilities, track progress, and resolve disputes efficiently.

Seek Legal Advice: Consult with legal professionals experienced in construction law to understand your rights and obligations under the contract and applicable laws. Legal advice can help navigate complex disputes and protect your interests in negotiations.

Managing Changes Effectively:

Change Order Procedures: Establish clear procedures for managing change orders, including documenting scope changes, estimating costs, and obtaining client approval. Clearly communicate the impact of changes on project timelines, budgets, and deliverables to manage expectations effectively.

Renegotiate Terms: In the event of significant scope changes or unforeseen circumstances, renegotiate contract terms with clients and stakeholders to reflect the revised project requirements, costs, and timelines.

Flexible Contract Provisions: Include flexible contract provisions that allow for reasonable adjustments to the scope, schedule, and budget in response to changes or unforeseen events. These provisions provide a framework for managing changes while maintaining project integrity and client satisfaction.

Regular Review and Adjustment: Continuously monitor project progress and reassess project plans, schedules, and budgets to identify potential changes and proactively address emerging issues. Regular review and adjustment ensure that projects remain aligned with client expectations and industry standards.

By adopting proactive communication strategies, implementing effective dispute resolution mechanisms, and managing changes efficiently, negotiators can navigate challenges and uncertainties in how-to home improvement projects with confidence and professionalism. In the subsequent chapters, we will explore additional negotiation techniques and best practices for achieving success in home improvement negotiations.

Chapter: Contract Negotiation in Engineering Construction

In this chapter, we will delve into the intricate world of contract negotiation within the engineering construction industry. Engineering construction projects encompass a wide range of infrastructure initiatives, from transportation and utilities to industrial facilities and commercial buildings. We will provide an overview of this dynamic industry, including major projects, regulatory considerations, and safety standards that shape contract negotiations.

Overview of the Engineering Construction Industry:

Major Projects: The engineering construction industry is characterized by large-scale projects that require specialized expertise and resources. These projects may include the construction of highways, bridges, airports, dams, power plants, and other critical infrastructure assets. Major projects often involve collaboration between government agencies, private developers, engineering firms, and construction contractors.

Regulatory Considerations: Engineering construction projects are subject to a myriad of regulations at the local, state, and federal levels. Regulatory requirements may include environmental permits, zoning approvals, building codes, and safety regulations. Negotiators must navigate these regulatory frameworks to ensure compliance and mitigate legal risks throughout the project lifecycle.

Safety Standards: Safety is paramount in engineering construction projects, given the inherent risks associated with heavy machinery, hazardous materials, and worksite conditions. Adherence to safety standards and protocols is non-negotiable, with strict regulations governing worker safety, equipment operation, and accident prevention. Negotiators must prioritize safety considerations in contract negotiations to protect workers, minimize liabilities, and maintain project continuity.

Key Considerations in Contract Negotiation:

Project Scope and Specifications: Negotiating the scope of work and project specifications is critical to defining the parameters of the engineering construction contract. This includes delineating the technical requirements, performance standards, and deliverables expected from the contractor. Clarity and specificity in defining the scope of work help minimize ambiguities and mitigate disputes during project execution.

Risk Allocation and Liability: Engineering construction projects involve inherent risks, ranging from design errors and material defects to unforeseen site conditions and force majeure events. Negotiators must carefully allocate risks and liabilities between parties through contractual provisions such as indemnification clauses, insurance requirements, and dispute resolution mechanisms. Balancing risk allocation ensures that parties bear responsibility commensurate with their ability to control and manage project risks.

Compliance with Regulatory Requirements: Contract negotiations must address compliance with applicable laws, regulations, and permits governing the engineering construction project. This includes obtaining necessary permits and approvals, adhering to environmental and safety standards, and complying with labor laws and prevailing wage requirements. Negotiators should ensure that contracts reflect these regulatory obligations and establish mechanisms for monitoring and enforcing compliance throughout the project lifecycle.

Project Schedule and Milestones: Negotiating project schedules and milestones is essential for ensuring timely completion of engineering construction projects. Negotiators should establish realistic timelines, account for potential delays and contingencies, and incorporate flexibility to accommodate changes or unforeseen circumstances. Clear communication and coordination between parties are essential for aligning project schedules and minimizing disruptions to project progress.

Conclusion:

Contract negotiation in the engineering construction industry requires a comprehensive understanding of project complexities, regulatory frameworks, and safety standards. By addressing key considerations such as project scope, risk allocation, regulatory compliance, and project scheduling, negotiators can facilitate successful contract agreements that mitigate risks, ensure compliance, and deliver value to all stakeholders involved. In the subsequent chapters, we will explore advanced negotiation strategies and best practices tailored to the unique challenges of engineering construction projects.

Chapter: Negotiating Complex Project Specifications

In engineering construction, negotiating complex project specifications is a cornerstone of successful contract agreements. These specifications encompass technical requirements, environmental considerations, and risk management strategies. In this chapter, we will explore the intricacies of negotiating these specifications, highlighting the challenges and strategies for achieving mutually beneficial outcomes.

Technical Requirements:

Defining Technical Standards: Negotiators must articulate clear technical standards and performance criteria for the engineering construction project. This includes specifying materials, methods, and design parameters to ensure compliance with industry best practices and project objectives.

Addressing Design Changes: Engineering projects often undergo design changes throughout the planning and execution phases. Negotiators should establish procedures for managing design modifications, including approval processes, documentation requirements, and cost implications. Flexibility in accommodating design changes while maintaining project integrity is essential for successful negotiations.

Navigating Technological Advancements: Rapid advancements in technology present opportunities and challenges for engineering construction projects. Negotiators should stay abreast of emerging technologies and innovations relevant to the project scope, evaluating their feasibility, cost-effectiveness, and potential impact on project outcomes.

Environmental Considerations:

Environmental Impact Assessments: Engineering construction projects are subject to environmental regulations and permitting requirements. Negotiators must conduct thorough environmental impact assessments to identify potential environmental risks and mitigation measures. This includes addressing issues such as habitat disturbance, air and water pollution, and ecosystem preservation.

Sustainable Practices: Incorporating sustainable practices into engineering construction projects is increasingly important for meeting regulatory requirements and addressing societal concerns. Negotiators should explore opportunities for integrating green technologies, energy-efficient designs, and environmentally friendly materials into project specifications.

Compliance with Regulatory Standards: Environmental regulations impose stringent requirements on engineering construction projects, including obtaining permits, conducting environmental assessments, and implementing mitigation measures. Negotiators should negotiate contracts that reflect these regulatory obligations and establish mechanisms for monitoring and ensuring compliance throughout the project lifecycle.

Risk Management:

Identifying Project Risks: Effective risk management begins with identifying and assessing project risks, including technical, financial, and external factors. Negotiators should conduct comprehensive risk assessments, considering potential hazards, uncertainties, and vulnerabilities that may impact project outcomes.

Allocating Risks and Liabilities: Negotiating risk allocation provisions is crucial for defining each party's responsibilities and liabilities in the event of project delays, cost overruns, or unforeseen events. Contractual mechanisms such as indemnification clauses, insurance requirements, and limitation of liability provisions help mitigate risks and protect parties from undue financial exposure.

Contingency Planning: Negotiators should develop contingency plans and risk mitigation strategies to address potential threats to project success. This may include establishing reserve funds, alternative procurement sources, or emergency response protocols to manage unexpected challenges and disruptions.

Conclusion:

Negotiating complex project specifications in engineering construction requires a multidisciplinary approach that balances technical requirements, environmental considerations, and risk management strategies. By addressing these aspects comprehensively and collaboratively, negotiators can craft contract agreements that promote project success, mitigate risks, and deliver value to all stakeholders involved. In the subsequent chapters, we will delve deeper into advanced negotiation techniques and case studies illustrating effective approaches to negotiating complex engineering construction projects.

Chapter: Balancing Cost and Quality in Engineering Construction

In engineering construction projects, balancing cost and quality is a delicate yet essential endeavor. This chapter will explore the strategies and considerations involved in achieving this balance, focusing on procurement methods, bidding processes, and value engineering techniques.

Procurement Methods:

Traditional Procurement: Under traditional procurement methods, the project owner selects a design team to develop detailed plans and specifications, followed by a competitive bidding process to select a contractor for construction. This method offers clarity in project scope but may limit innovation and collaboration among project stakeholders.

Design-Build: In the design-build approach, the project owner contracts with a single entity responsible for both design and construction. This method can streamline project delivery, reduce administrative burden, and promote collaboration between designers and builders. However, it may sacrifice design flexibility and owner control over project specifications.

Construction Management at Risk (CMAR): CMAR involves hiring a construction manager early in the project development phase to provide pre-construction services and input on project planning. The construction manager assumes the risk for construction costs and schedule performance, incentivizing cost-effective solutions and value engineering.

Bidding Processes:

Competitive Bidding: Competitive bidding invites multiple contractors to submit bids based on project specifications and requirements. This process promotes transparency, fosters competition, and helps ensure that project costs are competitive. However, it may prioritize cost over quality and innovation, leading to potential trade-offs in project outcomes.

Qualification-Based Selection (QBS): QBS emphasizes the qualifications and experience of contractors and design professionals rather than solely focusing on price. This approach allows project owners to prioritize quality, expertise, and past performance in contractor selection, mitigating the risk of selecting low-bid contractors with limited capabilities.

Negotiated Procurement: Negotiated procurement involves direct negotiations between the project owner and selected contractors or construction managers. This method allows for flexibility in negotiating project terms, scope, and pricing, enabling parties to tailor agreements to specific project requirements and objectives.

Value Engineering:

Identifying Cost-Saving Opportunities: Value engineering is a systematic process for identifying cost-saving opportunities without compromising project quality or performance. This may involve reevaluating design alternatives, materials selection, construction methods, and operational efficiencies to optimize project value.

Collaborative Problem-Solving: Value engineering encourages collaborative problem-solving among project stakeholders, including designers, engineers, contractors, and owners. By leveraging diverse perspectives and expertise, teams can identify innovative solutions, mitigate risks, and optimize project outcomes.

Lifecycle Cost Analysis: In addition to upfront construction costs, value engineering considers lifecycle costs, including operation, maintenance, and lifecycle performance. By evaluating long-term cost implications and benefits, project teams can make informed decisions that maximize project value and return on investment.

Conclusion:

Balancing cost and quality in engineering construction requires a strategic approach that considers procurement methods, bidding processes, and value engineering techniques. By selecting appropriate procurement methods, fostering collaboration among project stakeholders, and prioritizing value-based decision-making, project owners can achieve optimal outcomes that meet both budgetary constraints and quality standards. In the subsequent chapters, we will explore case studies and best practices illustrating successful approaches to balancing cost and quality in engineering construction projects.

Chapter: Mitigating Legal Risks and Ensuring Compliance

In the engineering construction industry, navigating legal complexities and ensuring compliance with industry standards are paramount for project success. This chapter will delve into strategies for mitigating legal risks and achieving compliance, covering contractual safeguards, regulatory considerations, and best practices for dispute resolution.

Contractual Safeguards:

Clear Contractual Language: Contracts should be drafted with clear, unambiguous language that outlines the rights, responsibilities, and obligations of all parties involved. Specify project scope, deliverables, timelines, payment terms, and dispute resolution mechanisms to minimize potential conflicts and ambiguities.

Risk Allocation Provisions: Allocate project risks among parties fairly and equitably through contractual provisions such as indemnification clauses, insurance requirements, and limitation of liability provisions. Clearly define each party's responsibilities and liabilities to mitigate legal exposure and protect against unforeseen contingencies.

Change Order Procedures: Establish clear procedures for managing change orders, including documentation requirements, approval processes, and cost implications. Define the scope of permissible changes, procedures for negotiating change orders, and mechanisms for addressing disputes arising from change order requests.

Regulatory Considerations:

Compliance with Building Codes: Ensure compliance with local, state, and federal building codes, regulations, and zoning ordinances applicable to engineering construction projects. Engage qualified professionals to conduct regulatory reviews, obtain necessary permits, and ensure that project designs and construction practices meet regulatory requirements.

Environmental Compliance: Environmental regulations impose stringent requirements on engineering construction projects, including environmental impact assessments, pollution prevention measures, and habitat conservation efforts. Implement environmental management plans, conduct regular environmental audits, and maintain compliance with applicable environmental laws and regulations.

Occupational Health and Safety: Prioritize worker safety by complying with occupational health and safety standards, regulations, and best practices. Implement safety protocols, provide appropriate training and personal protective equipment, and conduct regular safety inspections to prevent accidents, injuries, and regulatory violations.

Dispute Resolution Best Practices:

Negotiation and Mediation: Encourage open dialogue and negotiation between parties to resolve disputes amicably and efficiently. Mediation offers a collaborative and non-adversarial approach to dispute resolution, allowing parties to reach mutually acceptable solutions with the assistance of a neutral mediator.

Arbitration: Arbitration provides a private and expedited alternative to traditional litigation for resolving construction disputes. Parties agree to submit their claims to a neutral arbitrator, whose decision is binding and enforceable. Arbitration offers confidentiality, flexibility, and specialized expertise in construction matters.

Litigation Avoidance: Minimize the risk of litigation by proactively addressing potential disputes through effective communication, documentation, and risk management practices. Implement dispute resolution clauses in contracts, engage in early intervention and problem-solving, and seek legal advice to mitigate legal risks and avoid costly litigation.

Conclusion:

Mitigating legal risks and ensuring compliance with industry standards are critical considerations for engineering construction projects. By implementing contractual safeguards, adhering to regulatory requirements, and adopting best practices for dispute resolution, project stakeholders can minimize legal exposure, promote project success, and safeguard their interests throughout the project lifecycle. In the subsequent chapters, we will explore case studies and practical examples illustrating effective strategies for managing legal risks and achieving compliance in engineering construction projects.

Chapter: Contract Negotiation in Civil Construction

In this chapter, we will explore the nuances of contract negotiation within the civil construction sector. Civil construction encompasses a wide range of projects, including infrastructure development, public works, and zoning regulation compliance. Understanding the unique characteristics of the civil construction sector is essential for navigating contract negotiations effectively.

Understanding the Civil Construction Sector:

Infrastructure Projects: Civil construction projects often involve the development, maintenance, and expansion of infrastructure systems essential for societal functioning. This includes roads, bridges, highways, airports, railways, water supply networks, and wastewater treatment facilities. Negotiators must grasp the technical requirements, regulatory constraints, and funding mechanisms associated with infrastructure projects to negotiate contracts successfully.

Public Works: Many civil construction projects are commissioned by governmental entities or public agencies to serve the needs of communities and enhance public welfare. Public works projects may include the construction of schools, hospitals, parks, government buildings, and municipal utilities. Negotiating contracts for public works projects requires compliance with public procurement laws, competitive bidding processes, and accountability measures to ensure transparency and fairness.

Zoning Regulations: Zoning regulations govern land use and development activities within municipalities, dictating permissible land uses, building heights, setback requirements, and density restrictions. Civil construction projects must adhere to zoning regulations to obtain necessary permits and approvals from local authorities. Negotiators should be familiar with zoning ordinances and land use regulations applicable to specific project locations to address zoning-related issues in contract negotiations.

Key Considerations in Contract Negotiation:

Scope of Work: Negotiate the scope of work with clarity and specificity, delineating the tasks, deliverables, and performance standards expected from the contractor. Civil construction projects often involve complex engineering designs, site preparation, earthwork, grading, drainage, utilities installation, and pavement construction. Define the scope of work comprehensively to minimize ambiguity and mitigate disputes during project execution.

Regulatory Compliance: Ensure compliance with regulatory requirements, permits, and approvals necessary for civil construction projects. This includes environmental impact assessments, stormwater management plans, wetland mitigation measures, and archaeological surveys. Negotiate contracts that reflect regulatory obligations and establish mechanisms for monitoring and enforcing compliance throughout the project lifecycle.

Risk Management: Allocate project risks appropriately among parties through contractual provisions such as indemnification clauses, insurance requirements, and dispute resolution mechanisms. Civil construction projects involve inherent risks, including design errors, construction defects, weather delays, and unforeseen site conditions. Negotiators should negotiate risk allocation provisions that protect parties' interests while promoting project success and accountability.

Conclusion:

Contract negotiation in the civil construction sector requires a thorough understanding of infrastructure projects, public works, and zoning regulations. By addressing key considerations such as scope of work, regulatory compliance, and risk management, negotiators can facilitate successful contract agreements that meet project objectives, regulatory requirements, and stakeholder expectations. In the subsequent chapters, we will explore advanced negotiation strategies and case studies illustrating effective approaches to contract negotiation in civil construction projects.

Chapter: Negotiating with Government Agencies and Stakeholders

Engaging with government agencies and stakeholders is integral to the success of civil construction projects. This chapter explores the intricacies of negotiating with these entities, focusing on permits, land acquisition, and community engagement.

Permit Negotiation:

Understanding Permitting Processes: Civil construction projects require various permits and approvals from governmental bodies to ensure compliance with regulatory requirements. Negotiators must understand the permitting processes, including application procedures, timelines, and required documentation.

Navigating Regulatory Requirements: Each district has its own set of regulations governing land use, environmental protection, and construction activities. Negotiators must navigate these regulations effectively, identifying potential hurdles and developing strategies to address regulatory concerns during permit negotiations.

Building Relationships with Permitting Authorities: Establishing positive relationships with permitting authorities is crucial for expediting the permit approval process. Effective communication, transparency, and cooperation can help build trust and facilitate constructive negotiations, leading to timely issuance of permits.

Land Acquisition Negotiation:

Identifying Land Needs: Civil construction projects often require land acquisition for rights-of-way, easements, or project sites. Negotiators must assess land needs early in the project planning phase and develop strategies for acquiring the necessary parcels.

Negotiating Purchase Agreements: Negotiating land acquisition agreements involves discussions on purchase price, terms of sale, and conditions of transfer. Negotiators must conduct thorough due diligence, including property assessments, title searches, and environmental evaluations, to mitigate risks and ensure fair and equitable agreements.

Addressing Community Concerns: Land acquisition can impact adjacent property owners and communities. Negotiators should engage with affected stakeholders, address their concerns, and seek mutually beneficial solutions to minimize opposition and promote community acceptance of the project.

Community Engagement:

Stakeholder Outreach: Engaging with stakeholders, including residents, businesses, and community organizations, is essential for building support and addressing concerns related to civil construction projects. Open houses, public meetings, and stakeholder forums provide opportunities for dialogue and feedback.

Transparency and Communication: Transparent communication fosters trust and credibility with the community. Negotiators should provide accurate and timely information about project goals, impacts, and benefits, and solicit input from stakeholders to inform decision-making and project design.

Mitigating Impacts: Civil construction projects can disrupt local communities through noise, dust, traffic congestion, and other inconveniences. Negotiators should develop mitigation measures to minimize negative impacts, such as implementing construction phasing plans, scheduling work during off-peak hours, and providing alternative transportation options.

Conclusion:

Negotiating with government agencies and stakeholders is a complex but essential aspect of civil construction projects. By understanding permitting processes, navigating land acquisition negotiations, and engaging with the community proactively, negotiators can overcome challenges, build consensus, and facilitate successful project outcomes. In the subsequent chapters, we will explore case studies and best practices illustrating effective approaches to negotiating with government agencies and stakeholders in civil construction projects.

Chapter: Addressing Environmental Concerns

In the contemporary landscape of civil construction, addressing environmental concerns is not only a legal requirement but also a moral imperative. This chapter delves into strategies for integrating sustainability practices, implementing effective waste management, and spearheading conservation efforts within civil construction projects.

Sustainability Practices:

Incorporating Green Building Design: Embrace green building principles by incorporating sustainable design features such as energy-efficient systems, renewable materials, and passive design strategies. Negotiate with designers and engineers to integrate sustainability into project specifications, aiming to reduce energy consumption, minimize environmental impacts, and enhance occupant comfort.

Utilizing Renewable Energy Sources: Explore opportunities to incorporate renewable energy sources, such as solar, wind, or geothermal energy, into civil construction projects. Negotiate contracts with energy providers or invest in on-site renewable energy systems to reduce reliance on fossil fuels and lower carbon emissions.

Promoting Eco-Friendly Materials: Negotiate procurement contracts to prioritize the use of eco-friendly materials with low embodied energy, recycled content, and sustainable sourcing practices. Collaborate with suppliers to identify sustainable alternatives for construction materials, including recycled aggregates, reclaimed lumber, and low-emission adhesives.

Waste Management:

Implementing Construction Waste Reduction Strategies: Negotiate contracts with contractors and subcontractors to implement construction waste reduction strategies, such as waste segregation, recycling programs, and salvaging reusable materials. Set waste diversion targets and incentivize contractors to achieve recycling and reuse goals through contractual incentives or penalties.

Minimizing Environmental Pollution: Negotiate pollution prevention measures to minimize environmental contamination from construction activities, such as erosion control, sedimentation management, and stormwater runoff mitigation. Implement best management practices for construction site management to prevent soil erosion, protect water quality, and preserve natural habitats.

Managing Hazardous Materials Properly: Negotiate hazardous materials management plans to ensure safe handling, storage, and disposal of hazardous substances used in construction activities. Comply with regulatory requirements for hazardous materials management and provide training to personnel on proper handling procedures and emergency response protocols.

Conservation Efforts:

Protecting Natural Habitats: Negotiate environmental mitigation measures to protect natural habitats, biodiversity, and sensitive ecosystems affected by civil construction projects. Implement habitat restoration programs, wildlife crossings, and green infrastructure solutions to minimize ecological impacts and enhance environmental resilience.

Preserving Cultural Heritage: Negotiate cultural resource management plans to identify, document, and preserve archaeological and historic sites impacted by civil construction activities. Collaborate with archaeologists, historians, and Indigenous communities to develop mitigation strategies that safeguard cultural heritage resources while allowing project development to proceed.

Engaging in Community Stewardship: Foster community stewardship by engaging residents, stakeholders, and volunteers in conservation efforts associated with civil construction projects. Organize community clean-up events, tree planting initiatives, and educational programs to raise awareness about environmental issues and promote active participation in environmental stewardship.

Conclusion:

Addressing environmental concerns is integral to responsible and sustainable civil construction practices. By integrating sustainability principles, implementing effective waste management strategies, and spearheading conservation efforts, negotiators can minimize environmental impacts, promote resource efficiency, and contribute to the long-term resilience and well-being of communities and ecosystems affected by civil construction projects. In the subsequent chapters, we will explore case studies and practical examples illustrating successful approaches to addressing environmental concerns in civil construction projects.

Chapter: Navigating Public-Private Partnerships and Funding Arrangements

Public-private partnerships (PPPs) have emerged as a prominent model for delivering civil construction projects, offering a collaborative approach to project development, financing, and delivery. This chapter explores the intricacies of navigating PPPs and various funding arrangements in civil construction projects.

Understanding Public-Private Partnerships:

Collaborative Project Delivery: PPPs involve collaboration between public sector entities (such as government agencies or municipalities) and private sector partners (such as developers, investors, or contractors) to finance, develop, and operate civil construction projects. PPPs leverage the respective strengths of public and private sectors to deliver infrastructure projects efficiently and cost-effectively.

Types of PPP Models: PPPs encompass various models, including design-build-finance-operate (DBFO), design-build-operate (DBO), build-operate-transfer (BOT), and concession agreements. Each model defines the roles, responsibilities, and risks of public and private partners differently, depending on project characteristics, financing mechanisms, and desired outcomes.

Key Benefits of PPPs: PPPs offer several benefits, including access to private sector innovation, expertise, and resources; risk-sharing between public and private partners; accelerated project delivery; and performance-based incentives to achieve project objectives and service levels.

Navigating Funding Arrangements:

Identifying Funding Sources: Civil construction projects require significant capital investment for design, construction, and operation. Negotiators must identify and secure funding sources, including public funds, private investment, grants, loans, and bonds, to finance project costs and ensure financial viability.

Structuring Financing Packages: Negotiate financing packages that align with project requirements, risks, and cash flow projections. This may involve structuring debt financing, equity investments, or hybrid financing arrangements tailored to project needs and investor preferences.

Mitigating Financial Risks: Assess and mitigate financial risks associated with PPPs, including revenue volatility, cost overruns, regulatory changes, and market fluctuations. Negotiate contractual provisions, such as revenue guarantees, cost-sharing mechanisms, and financial reserves, to allocate risks appropriately and protect project viability.

Negotiating PPP Contracts:

Defining Project Objectives: Negotiate PPP contracts that clearly define project objectives, performance standards, and service levels expected from private sector partners. Establish key performance indicators (KPIs), milestones, and penalties for non-compliance to incentivize performance and ensure accountability.

Allocating Risks and Liabilities: Negotiate risk allocation provisions that balance risks between public and private partners based on their respective capabilities and risk tolerance. Define responsibilities for project financing, design, construction, operation, maintenance, and performance management to minimize disputes and ensure project success.

Ensuring Transparency and Accountability: Negotiate contracts that promote transparency, accountability, and public oversight in PPP transactions. Include provisions for public disclosure of project information, financial reporting, audit requirements, and mechanisms for public input and feedback to enhance transparency and public trust.

Conclusion:

Navigating public-private partnerships and funding arrangements requires careful planning, strategic negotiation, and collaboration between public and private sector stakeholders. By understanding PPP models, securing funding sources, and negotiating PPP contracts effectively, negotiators can leverage the strengths of both sectors to deliver successful civil construction projects that meet public needs, promote economic development, and enhance infrastructure resilience. In the subsequent chapters, we will explore case studies and best practices illustrating successful approaches to navigating PPPs and funding arrangements in civil construction projects.

Chapter: Contract Negotiation in Business

Business contract negotiation is a cornerstone of corporate transactions, facilitating commercial relationships, partnerships, and mergers that drive economic growth and innovation. This chapter provides an overview of business contract negotiation, focusing on commercial transactions, partnerships, and mergers within the corporate landscape.

Commercial Transactions:

Defining Commercial Contracts: Commercial contracts govern transactions between businesses, encompassing sales agreements, service contracts, procurement contracts, and distribution agreements. Negotiators must articulate the terms and conditions of these contracts, including pricing, delivery schedules, warranties, and dispute resolution mechanisms, to protect the interests of both parties.

Negotiating Terms and Conditions: Negotiate commercial contracts with clarity and specificity, addressing key terms such as payment terms, delivery terms, product specifications, intellectual property rights, and confidentiality provisions. Balance the interests of both parties to achieve mutually beneficial outcomes while minimizing legal risks and potential disputes.

Addressing Regulatory Compliance: Ensure compliance with regulatory requirements and industry standards applicable to commercial transactions, such as consumer protection laws, antitrust regulations, and data privacy regulations. Negotiate contracts that reflect regulatory obligations and establish mechanisms for monitoring and enforcing compliance to mitigate legal risks and ensure business continuity.

Partnerships:

Forming Strategic Partnerships: Strategic partnerships enable businesses to leverage complementary strengths, resources, and expertise to achieve common objectives and accelerate growth. Negotiators must identify strategic partners, assess partnership opportunities, and negotiate partnership agreements that align with business goals and maximize synergies.

Negotiating Partnership Agreements: Negotiate partnership agreements that define the rights, responsibilities, and obligations of each partner, including capital contributions, profit sharing arrangements, management structure, decision-making processes, and withdrawal plans. Establish clear governance mechanisms and dispute resolution procedures to promote transparency, accountability, and collaboration within the partnership.

Cultivating Collaborative Relationships: Foster collaborative relationships with partners through open communication, trust-building, and shared vision alignment. Establish channels for regular communication, performance monitoring, and feedback exchange to maintain productive partnerships and address challenges proactively.

Mergers:

Navigating Mergers and Acquisitions (M&A): Mergers and acquisitions involve the consolidation of businesses through the purchase, sale, or combination of assets, equity, or operations. Negotiators must navigate complex M&A transactions, including due diligence, valuation, negotiation, and integration planning, to achieve successful outcomes for all parties involved.

Negotiating Merger Agreements: Negotiate merger agreements that address key terms and conditions, including purchase price, deal structure, representations, and warranties, closing conditions, and post-closing obligations. Balance the interests of buyers, sellers, and other stakeholders to facilitate smooth transaction execution and mitigate transactional risks.

Managing Integration Challenges: Manage integration challenges post-merger, including cultural differences, organizational alignment, systems integration, and workforce transitions. Develop integration plans, communication strategies, and change management initiatives to facilitate seamless integration and maximize synergies while minimizing disruption to business operations.

Conclusion:

Contract negotiation in business plays a pivotal role in facilitating commercial transactions, partnerships, and mergers that drive corporate growth and success. By understanding the nuances of commercial contracts, partnerships, and M&A transactions, negotiators can navigate complex business landscapes, forge strategic alliances, and capitalize on opportunities for innovation and expansion. In the subsequent chapters, we will explore advanced negotiation techniques and case studies illustrating successful approaches to contract negotiation in various business contexts.

Chapter: Negotiating Sales and Procurement Contracts

Sales and procurement contracts form the backbone of business transactions, facilitating the exchange of goods and services between parties. This chapter explores the intricacies of negotiating these contracts, focusing on pricing models, delivery terms, and payment schedules to ensure mutually beneficial outcomes for buyers and sellers.

Understanding Sales Contracts:

Defining Sales Agreements: Sales contracts outline the terms and conditions governing the sale of goods or services between a seller and a buyer. These agreements specify product specifications, pricing, delivery terms, warranties, and dispute resolution mechanisms to formalize the transaction and protect the interests of both parties.

Negotiating Pricing Models: Negotiate pricing models that align with market dynamics, cost structures, and value propositions for both buyers and sellers. Common pricing models include fixed pricing, cost-plus pricing, volume-based pricing, and dynamic pricing strategies tailored to market demand, competitive landscape, and customer preferences.

Determining Delivery Terms: Negotiate delivery terms that define the responsibilities and obligations of both parties regarding product delivery, transportation, and risk transfer. Address key considerations such as shipping Incoterms, delivery locations, transit times, packaging requirements, and liability for loss or damage during transit to ensure timely and secure product delivery.

Navigating Procurement Contracts:

Identifying Procurement Needs: Procurement contracts govern the acquisition of goods, services, or works by organizations to fulfill operational needs or project requirements. Negotiators must identify procurement needs, assess supplier capabilities, and develop procurement strategies that optimize value, quality, and cost-effectiveness for the organization.

Negotiating Supplier Agreements: Negotiate supplier agreements that define the terms and conditions of procurement, including pricing, quantity discounts, lead times, quality standards, and service levels. Establish clear performance metrics, delivery schedules, and penalty clauses to ensure supplier accountability and mitigate supply chain risks.

Managing Payment Schedules: Negotiate payment schedules that balance cash flow requirements for both buyers and suppliers while ensuring timely and predictable payments for delivered goods or services. Discuss payment terms, invoicing procedures, milestone payments, and discounts for prompt payment to optimize cash management and strengthen supplier relationships.

Mitigating Risks and Ensuring Compliance:

Addressing Legal and Regulatory Compliance: Ensure compliance with legal and regulatory requirements applicable to sales and procurement contracts, such as consumer protection laws, contract law, commercial codes, and industry-specific regulations. Negotiate contracts that incorporate relevant legal provisions, warranties, and indemnification clauses to mitigate legal risks and ensure enforceability.

Managing Contractual Risks: Identify and mitigate contractual risks associated with sales and procurement contracts, including price fluctuations, supply chain disruptions, quality issues, and contractual breaches. Negotiate risk allocation provisions, force majeure clauses, and dispute resolution mechanisms to protect parties' interests and maintain contract performance.

Building Collaborative Relationships: Foster collaborative relationships with customers, suppliers, and stakeholders through transparent communication, trust-building, and shared value creation. Negotiate contracts that promote long-term partnerships, mutual respect, and continuous improvement to drive business growth, innovation, and sustainability.

Conclusion:

Negotiating sales and procurement contracts requires a strategic approach that balances the interests of buyers and sellers while addressing legal, financial, and operational considerations. By understanding pricing models, delivery terms, and payment schedules, negotiators can facilitate successful contract agreements that enhance value, minimize risks, and foster collaborative relationships between parties. In the subsequent chapters, we will explore advanced negotiation techniques and case studies illustrating successful approaches to sales and procurement contract negotiation in diverse business contexts.

Chapter: Drafting Effective Service Agreements

Service agreements are vital for governing relationships between service providers and clients, ensuring clarity, accountability, and mutual satisfaction. This chapter focuses on drafting effective service agreements, covering essential components such as performance metrics, service levels, and termination clauses to safeguard the interests of both parties.

Understanding Service Agreements:

Defining Service Scope: Service agreements delineate the scope of services to be provided by the service provider and the corresponding responsibilities of the client. Clearly define the services, deliverables, timelines, and expectations to establish a mutual understanding of the scope of work.

Negotiating Pricing and Payment Terms: Negotiate pricing structures and payment terms that align with the value of services rendered, budget constraints, and cash flow requirements of both parties. Specify payment schedules, invoicing procedures, and acceptable payment methods to facilitate prompt and transparent transactions.

Establishing Legal Framework: Service agreements establish the legal framework governing the relationship between the parties, including rights, obligations, remedies, and liabilities. Draft contracts with clear, unambiguous language that reflects the intentions of the parties and complies with applicable laws and regulations.

Performance Metrics and Service Levels:

Setting Performance Metrics: Define key performance indicators (KPIs) and performance metrics to measure the quality, timeliness, and effectiveness of services delivered. Identify measurable goals, benchmarks, and performance targets that align with the client's objectives and expectations.

Establishing Service Levels: Specify service levels that outline the minimum standards of performance expected from the service provider, including response times, resolution times, uptime guarantees, and service availability. Define service level agreements (SLAs) that quantify service performance and establish accountability mechanisms for meeting SLA targets.

Monitoring and Reporting: Implement monitoring and reporting mechanisms to track service performance, identify deviations from SLA targets, and address performance issues proactively. Establish regular reporting intervals, performance review meetings, and escalation procedures to maintain transparency and accountability.

Termination Clauses and Exit Strategies:

Drafting Termination Provisions: Include termination clauses that delineate the circumstances under which either party may terminate the agreement, such as breach of contract, non-performance, insolvency, or force majeure events. Specify notice periods, termination procedures, and consequences of termination to protect the interests of both parties.

Addressing Transition and Exit Strategies: Plan for contract termination or expiration by outlining transition and withdrawal plans to ensure continuity of services and mitigate disruptions to the client's operations. Develop contingency plans, data migration procedures, and knowledge transfer protocols to facilitate a smooth transition to alternative service providers or in-house operations.

Resolving Disputes and Claims: Establish dispute resolution mechanisms, such as mediation, arbitration, or litigation, to resolve disputes arising from contract termination or performance issues. Include provisions for dispute escalation, negotiation, and mediation to facilitate amicable resolution and minimize legal costs and reputational risks.

Conclusion:

Drafting effective service agreements requires careful consideration of performance metrics, service levels, and termination clauses to establish clear expectations, foster accountability, and mitigate risks. By incorporating these essential components into service agreements, service providers and clients can build trust, enhance collaboration, and achieve mutually beneficial outcomes. In the subsequent chapters, we will explore advanced negotiation techniques and case studies illustrating successful approaches to drafting and negotiating service agreements in diverse business contexts.

Chapter: Managing Intellectual Property Rights and Confidentiality Provisions

Intellectual property (IP) rights and confidentiality provisions are critical considerations in contracts, safeguarding valuable assets and proprietary information. This chapter delves into strategies for effectively managing IP rights and confidentiality provisions in contracts to protect the interests of parties involved.

Understanding Intellectual Property Rights:

Identifying Intellectual Property: Intellectual property encompasses various intangible assets, including patents, trademarks, copyrights, trade secrets, and proprietary information. Identify and classify relevant IP assets that are created, used, or exchanged during business transactions.

Protecting IP Rights: Negotiate contracts that specify ownership, licensing, and usage rights for IP assets to protect against unauthorized use, reproduction, or disclosure. Implement confidentiality measures, access controls, and technological safeguards to safeguard IP assets from infringement or misappropriation.

Managing IP Risks: Assess and mitigate risks associated with IP infringement, licensing disputes, and third-party claims through contractual provisions, indemnification clauses, and insurance coverage. Conduct due diligence on IP assets, including patent searches, trademark registrations, and copyright clearances, to mitigate legal risks and ensure compliance with intellectual property laws.

Implementing Confidentiality Provisions:

Defining Confidential Information: Define the scope of confidential information covered by confidentiality provisions, including trade secrets, proprietary data, customer lists, financial information, and technical know-how. Clearly identify confidential information through non-disclosure agreements (NDAs) or confidentiality clauses in contracts.

Establishing Confidentiality Obligations: Establish confidentiality obligations that restrict the disclosure, use, or reproduction of confidential information by parties to the contract. Specify the duration of confidentiality obligations, exceptions to confidentiality, and permitted disclosures to third parties to balance confidentiality with business needs.

Enforcing Confidentiality Provisions: Enforce confidentiality provisions through contractual remedies, including injunctive relief, damages, and termination rights for breaches of confidentiality. Implement security measures, such as encryption, password protection, and access controls, to prevent unauthorized access or disclosure of confidential information.

Balancing IP Rights and Confidentiality:

Negotiating IP Ownership and Licensing: Negotiate IP ownership and licensing arrangements that balance the interests of parties while preserving the value and integrity of IP assets. Define ownership rights, usage rights, and restrictions on the transfer or sublicensing of IP assets to protect against unauthorized use or exploitation.

Integrating IP and Confidentiality Provisions: Integrate IP rights and confidentiality provisions into contracts to provide comprehensive protection for proprietary information and intellectual property. Address overlap or conflicts between IP rights and confidentiality provisions to ensure consistency and enforceability in contract terms.

Mitigating Risks of IP Theft or Disclosure: Mitigate risks of IP theft or disclosure by implementing robust security measures, employee training programs, and contractual safeguards. Monitor compliance with confidentiality provisions, conduct periodic audits, and enforce contractual remedies to deter breaches and protect sensitive information from unauthorized access or disclosure.

Conclusion:

Managing intellectual property rights and confidentiality provisions is essential for protecting valuable assets and proprietary information in contracts. By understanding the nuances of IP rights, confidentiality obligations, and contractual remedies, parties can mitigate legal risks, safeguard confidential information, and preserve the value of intellectual property assets. In the subsequent chapters, we will explore advanced negotiation techniques and case studies illustrating successful approaches to managing IP rights and confidentiality provisions in contracts across diverse industries.

Chapter: Contract Negotiation in Administrative Law

Administrative law governs the relationship between individuals, businesses, and governmental agencies, regulating various aspects of public administration, including regulations, licensing requirements, and government oversight. This chapter provides an overview of administrative law principles relevant to contract negotiation, focusing on understanding regulations, navigating licensing requirements, and managing government oversight.

Understanding Administrative Law:

Regulatory Framework: Administrative law encompasses a complex framework of regulations, rules, and procedures established by government agencies to implement and enforce statutory laws. Regulations govern a wide range of activities, including environmental protection, consumer rights, labor standards, healthcare, and financial services.

Licensing and Permitting: Administrative agencies administer licensing and permitting systems to regulate professions, occupations, businesses, and activities that require governmental authorization. Licensing requirements vary by district and industry, requiring compliance with educational, training, experience, and examination criteria to obtain and maintain licenses.

Government Oversight: Administrative agencies exercise regulatory authority and government oversight to ensure compliance with laws, regulations, and administrative procedures. Agencies may conduct inspections, audits, investigations, and enforcement actions to monitor and enforce regulatory compliance, address violations, and protect public health, safety, and welfare.

Navigating Regulatory Compliance:

Identifying Applicable Regulations: Identify and understand applicable regulations that govern specific industries, activities, or transactions subject to administrative oversight. Conduct regulatory research, consult legal experts, and engage with regulatory agencies to interpret and apply regulations effectively in contract negotiation.

Compliance Assessments: Conduct compliance assessments to evaluate regulatory requirements, assess potential legal risks, and identify compliance gaps or deficiencies that may impact contract negotiations. Develop compliance strategies, risk mitigation measures, and regulatory compliance programs to address regulatory challenges and ensure adherence to legal requirements.

Regulatory Impact Analysis: Conduct regulatory impact analysis to assess the potential effects of regulations on contract negotiations, business operations, and financial performance. Evaluate regulatory costs, benefits, and implications for stakeholders to inform decision-making, risk management, and strategic planning in contract negotiations.

Navigating Licensing Requirements:

Understanding Licensing Processes: Understand licensing processes, procedures, and requirements established by regulatory agencies to obtain and maintain licenses. Familiarize yourself with application procedures, eligibility criteria, renewal requirements, and continuing education obligations for licensed professions, occupations, or businesses.

Negotiating Licensing Terms: Negotiate licensing terms and conditions with regulatory agencies to obtain favorable outcomes for license applicants. Advocate for reasonable licensing requirements, expedited processing, and streamlined procedures to minimize administrative burdens and facilitate timely licensure for individuals or businesses.

Managing License Compliance: Manage license compliance obligations by implementing systems, processes, and controls to track license statuses, renewal deadlines, and continuing education requirements. Develop compliance protocols, recordkeeping practices, and audit procedures to maintain license eligibility, prevent lapses, and avoid regulatory sanctions or penalties.

Managing Government Oversight:

Engaging with Regulatory Agencies: Engage proactively with regulatory agencies to build constructive relationships, foster open communication, and address regulatory concerns in contract negotiations. Seek guidance, clarification, and feedback from regulatory authorities to ensure compliance with regulatory requirements and resolve regulatory issues effectively.

Responding to Regulatory Inquiries: Respond promptly and professionally to regulatory inquiries, requests for information, or enforcement actions initiated by regulatory agencies. Cooperate with regulatory investigations, provide accurate and timely responses, and implement corrective actions to address compliance deficiencies and mitigate regulatory risks.

Navigating Enforcement Actions: Navigate enforcement actions, administrative proceedings, or regulatory disputes initiated by regulatory agencies through effective advocacy, negotiation, and resolution strategies. Engage legal counsel, regulatory experts, and industry representatives to represent your interests, challenge regulatory actions, and seek favorable outcomes in administrative proceedings.

Conclusion:

Contract negotiation in administrative law requires a comprehensive understanding of regulations, licensing requirements, and government oversight mechanisms that govern business activities and transactions. By navigating regulatory compliance, licensing requirements, and government oversight effectively, parties can negotiate contracts that comply with legal requirements, mitigate regulatory risks, and promote regulatory compliance in administrative law contexts. In the subsequent chapters, we will explore advanced negotiation techniques and case studies illustrating successful approaches to contract negotiation in administrative law.

Chapter: Negotiating Contracts with Regulatory Bodies

Negotiating contracts with regulatory bodies is a unique aspect of administrative law, involving interactions with governmental agencies responsible for enforcing regulations and overseeing compliance. This chapter explores the complexities of negotiating such contracts, focusing on compliance audits, inspections, and enforcement actions to ensure regulatory compliance and mitigate legal risks.

Understanding Regulatory Bodies:

Roles and Responsibilities: Regulatory bodies are governmental agencies tasked with creating, implementing, and enforcing regulations to protect public health, safety, and welfare. These agencies oversee compliance with statutory laws, administrative rules, and licensing requirements across various industries, sectors, and districts.

Regulatory Enforcement: Regulatory bodies conduct compliance audits, inspections, and enforcement actions to monitor and enforce regulatory compliance. They have the authority to investigate violations, issue citations, impose penalties, and take enforcement actions to address non-compliance and protect public interests.

Contractual Relationships: Negotiating contracts with regulatory bodies involves establishing formal agreements, consent decrees, or settlement agreements to address regulatory concerns, resolve compliance issues, and establish mutually acceptable terms and conditions for regulatory compliance and enforcement.

Navigating Compliance Audits:

Preparing for Audits: Prepare for compliance audits by conducting internal assessments, reviewing regulatory requirements, and gathering relevant documentation to demonstrate compliance with applicable laws and regulations. Anticipate areas of scrutiny, identify potential compliance gaps, and develop corrective action plans to address deficiencies proactively.

Cooperating with Auditors: Cooperate with regulatory auditors during compliance audits, providing access to records, facilities, and personnel as requested. Respond promptly to audit findings, address auditor inquiries, and provide explanations or clarifications to resolve compliance issues and facilitate audit completion.

Negotiating Audit Findings: Negotiate audit findings with regulatory bodies through informal discussions, formal meetings, or written responses to address discrepancies, clarify interpretations, and resolve disputes over regulatory compliance. Seek opportunities for remediation, corrective actions, or compliance enhancements to mitigate regulatory risks and prevent future violations.

Managing Inspections:

Responding to Inspection Notices: Respond promptly to inspection notices from regulatory bodies, coordinating site visits, and preparing for inspections to ensure compliance with regulatory requirements. Designate responsible personnel to accompany inspectors, facilitate inspections, and address inspection findings in real-time.

Addressing Inspection Findings: Address inspection findings promptly and thoroughly, documenting corrective actions, implementing remedial measures, and providing evidence of compliance to regulatory inspectors. Collaborate with regulatory agencies to resolve compliance issues, implement corrective actions, and prevent recurrence of violations.

Negotiating Compliance Plans: Negotiate compliance plans with regulatory bodies to address persistent compliance issues, develop corrective action plans, and establish milestones for achieving compliance objectives. Collaborate with regulatory agencies to negotiate mutually acceptable terms, deadlines, and performance metrics to ensure effective compliance management.

Responding to Enforcement Actions:

Understanding Enforcement Actions: Respond to enforcement actions initiated by regulatory bodies, including warning letters, notices of violation, administrative citations, and cease-and-desist orders. Understand the basis for enforcement actions, evaluate legal options, and develop strategies to challenge, settle, or mitigate regulatory liabilities and penalties.

Negotiating Settlements: Negotiate settlements or consent decrees with regulatory bodies to resolve enforcement actions, mitigate penalties, and achieve compliance objectives. Engage in settlement discussions, negotiate favorable terms, and document agreements to address regulatory concerns and avoid protracted legal disputes.

Implementing Corrective Actions: Implement corrective actions required by regulatory settlements or consent decrees, monitor compliance with agreed-upon terms, and provide periodic progress reports to regulatory agencies. Collaborate with regulatory inspectors to verify compliance, address outstanding issues, and demonstrate commitment to regulatory compliance and enforcement.

Conclusion:

Negotiating contracts with regulatory bodies requires strategic planning, proactive engagement, and effective communication to ensure regulatory compliance and mitigate legal risks. By navigating compliance audits, inspections, and enforcement actions effectively, parties can negotiate mutually beneficial agreements that address regulatory concerns, resolve compliance issues, and promote a culture of regulatory compliance in administrative law contexts. In the subsequent chapters, we will explore advanced negotiation techniques and case studies illustrating successful approaches to negotiating contracts with regulatory bodies.

Chapter: Addressing Legal Challenges and Regulatory Uncertainties

In the landscape of administrative law, legal challenges and regulatory uncertainties are inevitable. This chapter delves into strategies for addressing these challenges, focusing on dispute resolution mechanisms and administrative appeals to navigate through regulatory complexities and legal uncertainties effectively.

Understanding Legal Challenges:

Nature of Legal Challenges: Legal challenges in administrative law arise from regulatory enforcement actions, compliance disputes, administrative decisions, and interpretation of laws and regulations. These challenges can lead to conflicts, disputes, and uncertainties that require resolution through legal mechanisms.

Sources of Regulatory Uncertainties: Regulatory uncertainties stem from evolving laws, ambiguous regulations, conflicting interpretations, and changing enforcement priorities. Regulatory uncertainties can create compliance risks, operational challenges, and legal liabilities for businesses, individuals, and governmental agencies.

Impact on Contract Negotiations: Legal challenges and regulatory uncertainties can impact contract negotiations by introducing legal risks, complicating compliance requirements, and affecting parties' ability to reach mutually acceptable terms and conditions. Addressing these challenges requires proactive strategies and effective dispute resolution mechanisms.

Navigating Dispute Resolution Mechanisms:

Choosing Dispute Resolution Mechanisms: Select appropriate dispute resolution mechanisms, such as negotiation, mediation, arbitration, or litigation, based on the nature, complexity, and urgency of legal challenges. Evaluate the advantages, disadvantages, and costs associated with each mechanism to determine the most suitable approach.

Negotiating Settlements: Explore settlement negotiations as a primary means of resolving legal challenges and regulatory uncertainties. Engage in collaborative discussions, identify common interests, and negotiate mutually acceptable solutions to mitigate legal risks, preserve business relationships, and avoid protracted legal disputes.

Seeking Mediation or Arbitration: Consider mediation or arbitration as alternative dispute resolution mechanisms to resolve legal challenges efficiently, impartially, and cost-effectively. Engage neutral mediators or arbitrators to facilitate constructive dialogue, clarify legal issues, and facilitate consensus-building among parties to achieve amicable resolutions.

Navigating Administrative Appeals:

Understanding Administrative Appeals: Administrative appeals provide recourse for challenging adverse decisions, actions, or rulings issued by governmental agencies through administrative procedures. Administrative appeals may involve formal hearings, administrative reviews, or appellate reviews before administrative tribunals or judicial courts.

Filing Appeals Timely: File administrative appeals promptly within statutory deadlines, procedural requirements, and jurisdictional limitations established by administrative regulations or statutes. Comply with filing procedures, submission requirements, and notice provisions to preserve appeal rights and avoid procedural pitfalls.

Presenting Appellate Arguments: Present persuasive appellate arguments supported by relevant evidence, legal precedents, and statutory interpretations to substantiate grounds for appeal. Articulate legal theories, factual assertions, and procedural objections clearly and convincingly to demonstrate errors, abuses of discretion, or violations of law warranting appellate relief.

Managing Regulatory Compliance During Disputes:

Continuing Regulatory Compliance: Maintain regulatory compliance obligations during legal challenges and dispute resolution proceedings to mitigate legal risks, preserve business operations, and uphold regulatory responsibilities. Adhere to statutory requirements, regulatory standards, and enforcement directives while navigating legal disputes and administrative appeals.

Cooperating with Regulatory Authorities: Cooperate with regulatory authorities during legal challenges, dispute resolution proceedings, and administrative appeals to demonstrate good faith, transparency, and commitment to regulatory compliance. Engage in open dialogue, provide requested information, and comply with regulatory requests to facilitate resolution and minimize regulatory scrutiny.

Mitigating Compliance Risks: Mitigate compliance risks associated with legal challenges and regulatory uncertainties by implementing risk management strategies, compliance programs, and internal controls. Monitor regulatory developments, assess compliance risks, and adapt compliance measures to address emerging legal challenges and regulatory uncertainties effectively.

Conclusion:

Addressing legal challenges and regulatory uncertainties in administrative law requires a multifaceted approach that incorporates dispute resolution mechanisms, administrative appeals, and ongoing regulatory compliance efforts. By navigating legal challenges proactively, engaging in effective dispute resolution strategies, and maintaining regulatory compliance during disputes, parties can resolve conflicts, mitigate legal risks, and achieve favorable outcomes in administrative law contexts. In the subsequent chapters, we will explore advanced negotiation techniques and case studies illustrating successful approaches to addressing legal challenges and regulatory uncertainties.

Chapter: Ensuring Transparency and Accountability in Government Contracts and Public-Private Partnerships

Transparency and accountability are fundamental principles in government contracts and public-private partnerships (PPPs), ensuring public trust, integrity, and effective governance. This chapter explores strategies for fostering transparency and accountability throughout the lifecycle of government contracts and PPPs to promote ethical conduct, mitigate corruption risks, and enhance public value.

Importance of Transparency and Accountability:

Enhancing Public Trust: Transparency and accountability promote public trust by providing stakeholders with access to information, decision-making processes, and performance outcomes related to government contracts and PPPs. Transparency fosters openness, honesty, and integrity in public administration, enhancing credibility and legitimacy.

Preventing Corruption: Transparency and accountability serve as essential safeguards against corruption, fraud, and abuse of power in government contracts and PPPs. Openness in procurement processes, disclosure of financial transactions, and oversight mechanisms deter corrupt practices and promote fair competition, ensuring taxpayer dollars are used efficiently and ethically.

Facilitating Effective Governance: Transparency and accountability facilitate effective governance by promoting informed decision-making, citizen engagement, and oversight of government actions. Transparency enables stakeholders to monitor government performance, evaluate policy outcomes, and hold public officials accountable for their actions, fostering democratic accountability and responsiveness.

Promoting Transparency in Government Contracts:

Open Procurement Processes: Implement transparent procurement processes that provide equal opportunities for all qualified vendors, contractors, and suppliers to compete for government contracts. Publish procurement notices, solicit bids openly, and disclose evaluation criteria to promote competition, fairness, and transparency in contract award decisions.

Public Access to Information: Provide public access to contract-related information, documents, and performance data through centralized portals, databases, or public records requests. Publish contract awards, terms, and performance reports online to enhance transparency, accountability, and public scrutiny of government spending and contracting activities.

Stakeholder Engagement: Engage stakeholders, including civil society organizations, advocacy groups, and affected communities, in contract negotiations, decision-making processes, and project oversight mechanisms. Solicit public input, feedback, and participation to enhance transparency, legitimacy, and public confidence in government contracts and PPPs.

Ensuring Accountability in Public-Private Partnerships:

Robust Governance Structures: Establish robust governance structures, oversight mechanisms, and accountability frameworks for PPPs to ensure compliance with contractual obligations, regulatory requirements, and public policy objectives. Designate accountable authorities, clarify roles and responsibilities, and establish reporting mechanisms to monitor PPP performance and mitigate risks.

Performance Monitoring and Evaluation: Implement performance monitoring and evaluation systems to assess PPP outcomes, measure service delivery, and track project performance against predefined targets and benchmarks. Conduct regular audits, reviews, and evaluations to identify gaps, address deficiencies, and hold PPP stakeholders accountable for meeting contractual obligations.

Transparency in Financial Transactions: Ensure transparency in financial transactions, revenue streams, and cost-sharing arrangements within PPP contracts. Disclose project financing details, funding sources, and financial projections to stakeholders to enhance accountability, mitigate financial risks, and ensure responsible stewardship of public resources.

Enforcing Compliance and Oversight:

Compliance Monitoring: Strengthen compliance monitoring mechanisms, enforcement procedures, and sanctions for PPPs to deter non-compliance, enforce contractual provisions, and address performance failures. Conduct inspections, audits, and compliance reviews to detect violations, enforce corrective actions, and hold PPP stakeholders accountable for their contractual obligations.

Independent Audits and Reviews: Commission independent audits, performance evaluations, and due diligence reviews of government contracts and PPPs to assess project effectiveness, efficiency, and compliance with legal and regulatory requirements. Engage external auditors, oversight bodies, and independent experts to provide impartial assessments and recommendations for improvement.

Ombudsman and Grievance Mechanisms: Establish ombudsman offices, grievance mechanisms, or dispute resolution processes to address complaints, grievances, and disputes arising from government contracts and PPPs. Provide channels for stakeholders to report concerns, seek redress, and access impartial mediation or arbitration to resolve conflicts and promote fairness in contract implementation.

Conclusion:

Ensuring transparency and accountability in government contracts and public-private partnerships is essential for promoting ethical conduct, preventing corruption, and enhancing public confidence in government actions. By promoting transparency, fostering accountability, and strengthening oversight mechanisms throughout the contract lifecycle, governments can uphold integrity, safeguard public interests, and deliver value for money in public procurement and service delivery. In the subsequent chapters, we will explore advanced negotiation techniques and case studies illustrating successful approaches to ensuring transparency and accountability in government contracts and PPPs.

Conclusion: Mastering the Art of Contract Negotiation

Throughout this book, we've delved into the intricacies of contract negotiation across various industries and contexts, exploring key principles, strategies, and best practices to empower negotiators in achieving successful outcomes. As we conclude our journey, let's recap the fundamental principles and strategies discussed in this comprehensive guide:

Preparation is Key: Success in contract negotiation begins with thorough preparation. Research the parties involved, understand the industry dynamics, and clarify your objectives and priorities before entering negotiations.

Communication is Critical: Effective communication is the cornerstone of successful negotiation. Practice active listening, ask probing questions, and build rapport with counterparties to foster trust, understanding, and collaboration.

Focus on Interests, not Positions: Shift the focus from positional bargaining to interest-based negotiation. Identify underlying interests, needs, and concerns of all parties and explore creative solutions that satisfy mutual interests and create value.

Seek Win-Win Solutions: Strive for mutually beneficial agreements that maximize value for all parties involved. Explore integrative negotiation strategies, such as expanding the pie, joint problem-solving, and trade-offs, to create value and build long-term relationships.

Understand Legal and Regulatory Frameworks: Familiarize yourself with legal and regulatory requirements relevant to your industry and district. Ensure compliance with laws, regulations, and contractual obligations to mitigate legal risks and uphold integrity in contract negotiation.

Customize Strategies for Each Industry: Recognize the unique dynamics and challenges of contract negotiation in different industries, such as construction, business, administrative law, and public-private partnerships. Tailor negotiation strategies and tactics to address industry-specific issues and priorities effectively.

Embrace Transparency and Accountability: Foster transparency, accountability, and good governance principles in contract negotiation and implementation. Promote open communication, disclose relevant information, and establish mechanisms for oversight and compliance to build trust and credibility.

Navigate Disputes and Challenges: Anticipate and address disputes, challenges, and uncertainties that may arise during contract negotiation and implementation. Develop strategies for resolving conflicts, mitigating risks, and enforcing contractual rights to protect your interests and preserve relationships.

Continuous Learning and Improvement: Contract negotiation is a dynamic and iterative process that requires continuous learning and improvement. Reflect on your negotiation experiences, seek feedback from peers and mentors, and adapt your strategies based on lessons learned and emerging best practices.

Ethics and Professionalism: Conduct negotiations with integrity, honesty, and professionalism. Uphold ethical standards, respect the interests and rights of all parties, and adhere to principles of fairness, honesty, and respect to build trust and credibility in negotiations.

In mastering the art of contract negotiation, remember that success is not measured solely by securing favorable terms but by building relationships, creating value, and achieving outcomes that meet the needs and objectives of all parties involved. By embracing these principles and strategies, negotiators can navigate complex negotiations with confidence, skill, and integrity, fostering mutually beneficial agreements and sustainable partnerships.

As you embark on your negotiation journey, may this guide serve as a valuable resource and source of inspiration in mastering the art of contract negotiation across diverse industries and contexts. Best wishes for your future negotiations and endeavors!

Chapter: Reflecting on Personal Growth and Professional Development in Negotiation Skills

As we conclude our exploration of contract negotiation, it's crucial to take a moment to reflect on our personal growth and professional development in negotiation skills. Negotiation is not just about mastering techniques; it's also about continual learning, self-awareness, and growth. In this chapter, we'll reflect on our journey, celebrate achievements, and identify areas for further improvement in negotiation skills.

Acknowledge Progress:

Begin by acknowledging the progress you've made in negotiation skills. Reflect on past negotiations and recognize the milestones you've achieved, whether it's closing a significant deal, resolving a challenging dispute, or building stronger relationships with counterparties. Celebrate your successes and the lessons learned along the way.

Identify Strengths:

Reflect on your strengths as a negotiator. Consider the skills and qualities that have contributed to your success in negotiation, such as effective communication, strategic thinking, empathy, or resilience. Acknowledge these strengths and leverage them in future negotiations to achieve favorable outcomes and build confidence as a negotiator.

Recognize Areas for Improvement:

Be honest with yourself about areas where you can improve your negotiation skills. Reflect on past challenges, mistakes, or missed opportunities in negotiation and identify patterns or recurring issues that need attention. Whether it's improving active listening, managing emotions, or sharpening analytical skills, acknowledge areas for growth and commit to continuous improvement.

Seek Feedback:

Seek feedback from colleagues, mentors, or trusted advisors on your negotiation skills. Ask for honest assessments of your strengths and weaknesses in negotiation and solicit constructive feedback on areas for improvement. Embrace feedback as a valuable opportunity for learning and growth and use it to refine your negotiation approach and enhance effectiveness.

Set Learning Goals:

Set learning goals to further develop your negotiation skills. Identify specific areas or techniques you want to improve, such as negotiation strategy development, conflict resolution, or cross-cultural negotiation. Set SMART (Specific, Measurable, Achievable, Relevant, Time-bound) goals to guide your learning journey and track progress over time.

Invest in Professional Development:

Invest in professional development opportunities to expand your negotiation skills and knowledge. Attend negotiation workshops, seminars, or training programs to gain insights from experts, learn new techniques, and network with peers in the field. Consider pursuing certifications or advanced degrees in negotiation or related disciplines to deepen your expertise.

Practice Self-Reflection:

Cultivate a habit of self-reflection in your negotiation practice. Regularly review your negotiation experiences, analyze outcomes, and identify lessons learned. Journaling, meditation, or debriefing after negotiations can help you gain insights into your negotiation style, strengths, and areas for improvement.

Embrace Growth Mindset:

Embrace a growth mindset in your negotiation journey. View challenges as opportunities for learning and growth, rather than setbacks or failures. Embrace experimentation, adaptability, and resilience in negotiation, and approach each negotiation with a mindset of curiosity, openness, and continuous improvement.

Conclusion:

Reflecting on personal growth and professional development in negotiation skills is essential for becoming a more effective and confident negotiator. By acknowledging progress, identifying strengths, recognizing areas for improvement, seeking feedback, setting learning goals, investing in professional development, practicing self-reflection, and embracing a growth mindset, you can continue to evolve and excel in negotiation. As you continue your negotiation journey, may you find fulfillment in the pursuit of mastery and the pursuit of excellence in negotiation.

Chapter: Embracing Lifelong Learning and Adaptation in the Industry

As we conclude our journey through the realm of contract negotiation, it's crucial to recognize that learning and adaptation are ongoing processes in the ever-evolving landscape of industries. In this chapter, we'll explore the importance of embracing lifelong learning and adaptation to navigate changes effectively and stay ahead in the industry.

Embracing Lifelong Learning:

The field of contract negotiation is dynamic, with contemporary trends, technologies, and regulations continually shaping the landscape. Embracing lifelong learning is essential to staying updated with industry developments, emerging best practices, and evolving negotiation strategies. Commit to continuous learning through reading industry publications, attending seminars, participating in workshops, and seeking out mentorship opportunities to expand your knowledge and skills in negotiation.

Adapting to Changes in the Industry:

The industry of contract negotiation is not static; it undergoes constant changes driven by factors such as technological advancements, regulatory reforms, economic shifts, and market trends. To thrive in this dynamic environment, it's crucial to embrace adaptability and flexibility. Be open to change, willing to experiment with innovative approaches, and agile in responding to emerging challenges and opportunities in negotiation. Cultivate a mindset of adaptability, resilience, and innovation to navigate changes effectively and stay competitive in the industry.

Anticipating Future Trends:

Stay proactive in anticipating future trends and developments in the industry of contract negotiation. Monitor market trends, industry forecasts, and emerging technologies to identify potential opportunities and challenges on the horizon. Anticipate changes in consumer preferences, regulatory requirements, and competitive dynamics that may impact negotiation strategies and outcomes. By staying ahead of the curve, you can position yourself strategically and adapt your approach to capitalize on emerging trends and stay relevant in the industry.

Fostering Collaboration and Knowledge Sharing:

Recognize the value of collaboration and knowledge sharing in the industry of contract negotiation. Engage with peers, colleagues, and industry experts to exchange insights, share experiences, and learn from each other's perspectives. Participate in industry forums, networking events, and professional associations to foster collaborative learning and collective intelligence in negotiation. By leveraging the collective wisdom of the industry, you can gain fresh perspectives, innovative ideas, and practical strategies to enhance your negotiation practice.

Conclusion:

In the fast-paced and dynamic industry of contract negotiation, embracing lifelong learning and adaptation is essential for staying relevant, resilient, and effective. By committing to continuous learning, embracing adaptability, anticipating future trends, and fostering collaboration and knowledge sharing, you can navigate changes effectively, seize opportunities, and thrive in the ever-evolving landscape of negotiation. As you continue your journey in the industry, may you approach each new challenge with curiosity, enthusiasm, and a commitment to lifelong growth and adaptation.

Chapter: Upholding Ethical Conduct, Mutual Respect, and Win-Win Outcomes in Contract Negotiation

As we conclude our exploration of contract negotiation, it's essential to underscore the significance of ethical conduct, mutual respect, and win-win outcomes in fostering trust, integrity, and sustainable partnerships. In this concluding chapter, we will reflect on the importance of these principles and their profound impact on the negotiation process and outcomes.

Ethical Conduct:

Ethical conduct forms the foundation of principled negotiation, guiding negotiators to uphold honesty, integrity, and fairness in their interactions. Acting ethically means adhering to moral principles, respecting the rights and interests of all parties, and avoiding deceitful or manipulative tactics. In contract negotiation, ethical conduct builds trust, credibility, and goodwill among negotiators, fostering constructive dialogue and fostering long-term relationships based on mutual respect and integrity.

Mutual Respect:

Mutual respect is fundamental to productive negotiation, recognizing the dignity, autonomy, and perspectives of all parties involved. Respecting each other's viewpoints, interests, and needs cultivates an atmosphere of trust, openness, and collaboration, enabling negotiators to engage in constructive dialogue and explore creative solutions. In contract negotiation, mutual respect fosters empathy, understanding, and effective communication, paving the way for mutually beneficial agreements and sustainable partnerships built on trust and mutual benefit.

Win-Win Outcomes:

Seeking win-win outcomes is the hallmark of successful negotiation, where all parties derive value and satisfaction from the agreement reached. Win-win negotiation focuses on creating value, expanding the pie, and finding creative solutions that address the interests and needs of all parties involved. By prioritizing collaboration, problem-solving, and compromise, negotiators can achieve outcomes that maximize mutual gains, foster goodwill, and preserve relationships for the long term.

The Interplay of Ethical Conduct, Mutual Respect, and Win-Win Outcomes:

Ethical conduct, mutual respect, and win-win outcomes are interconnected principles that reinforce each other in the negotiation process. Acting ethically builds trust and respect among negotiators, creating a conducive environment for exploring win-win solutions. Mutual respect fosters empathy and understanding, facilitating collaboration and consensus-building towards win-win outcomes. Win-win outcomes, in turn, reinforce ethical conduct and mutual respect by demonstrating a commitment to fairness, integrity, and mutual benefit in negotiation.

Final Thoughts:

In the complex and dynamic landscape of contract negotiation, upholding ethical conduct, mutual respect, and win-win outcomes is essential for building trust, fostering collaboration, and achieving sustainable success. By embracing these principles, negotiators can navigate challenges, resolve conflicts, and forge enduring partnerships based on trust, integrity, and shared value. As you continue your negotiation journey, may you always strive to act ethically, treat others with respect, and pursue win-win outcomes that create value and promote mutual prosperity in contract negotiation.

www.ingramcontent.com/pod-product-compliance
Lightning Source LLC
Chambersburg PA
CBHW050328230526
45471CB00005B/2397